新雅・點讀樂園

新雅幼兒互動點讀圖典

新雅文化事業有限公司

www.sunya.com.hk

目錄

使用說明 —————— 3

My Body
我的身體 —————— 6

My Family
我的家庭 —————— 8

In the Living Room
客廳裏 —————— 10

In the Kitchen
廚房中 —————— 12

Bedroom and Bathroom
睡房和浴室 —————— 14

My School
我的學校 —————— 16

I Can...
我會…… —————— 18

My Toys
我的玩具 —————— 20

Food and Drinks
食物和飲品 —————— 22

Vegetables and Fruit
蔬菜和水果 —————— 24

Things to Wear
日常衣物 —————— 26

Jobs
不同的職業 —————— 28

Things That Go
交通工具 —————— 30

In the City
城市裏 —————— 32

At the Park
公園裏 —————— 34

My Pets
我的寵物 —————— 36

On the Farm
農場裏 —————— 37

At the Zoo
動物園裏 —————— 38

Months, Days and Time
月份、日和時間 —————— 40

Seasons and Weather
季節和天氣 —————— 42

Numbers
數字 —————— 44

Colours and Shapes
顏色和形狀 —————— 45

Adjective
形容詞 —————— 46

Opposite
相反詞 —————— 48

Festivals
節日 —————— 50

Musical Instruments
樂器 —————— 52

Love the Earth
愛護地球 —————— 53

Space and Landform
太空和地貌 —————— 54

Spelling Game
拼字遊戲 —————— 56

Dictionary
小字典 —————— 68

使用說明

　　《新雅幼兒互動點讀圖典》配合新雅點讀筆使用，讓幼兒愉快學習600個生活必備的常用字。幼兒不但可點讀字詞，聆聽英、粵、普的發音，還可以玩互動主題遊戲和拼字遊戲，提升幼兒的認讀能力、拼字能力和自學能力。此外，本圖典還可配合拼字遊戲卡使用。幼兒透過動手拼出600個英文生字，能加深他們的學習記憶，富趣味又能鞏固所學。

　　本圖典屬新雅點讀樂園產品之一，想了解更多新雅的點讀產品，請瀏覽新雅網頁 (www.sunya.com.hk) 或掃描右邊的 QR code 進入 新雅・點讀樂園

1 點讀筆的基本功能

USB 插口
使用附送的 USB 線連接電腦充電或傳輸檔案。

喇叭

耳機孔
可連接耳機使用

音量鍵
短按音量⊕或⊖鍵，調節音量大小。

電源開關
長按 3 秒，開啟電源。再長按 3 秒，則關閉電源。待機 5 分鐘後會有語音提示是否繼續操作，30 秒內沒有繼續操作則自動關機。

指示燈
啟動時顯示為藍色。充電時變為紅色，充滿電後會熄滅，同時點讀筆自動斷電，以作保護。

光學感應筆頭
用筆頭接觸圖典上的圖示、文字或圖畫，點讀筆會播放相應的內容。

⭐ 點讀筆的其他功能請詳見《新雅點讀筆說明書》。

② 如何讓孩子愉快學習 600 個常用字

❶ 先點選封面，啟動點讀圖典內容的功能，然後點選頁面上的字詞或圖畫，點讀筆便能讀出相關內容。

❷ 點讀筆的預設語言為粵語。如想切換語言，可用點讀筆點選圖示，當再次點選頁面時，點讀筆便會以所選的語言讀出相關內容。例如選擇以英語播放時：

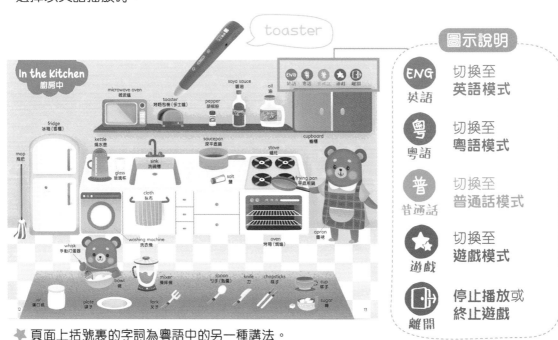

🌟 頁面上括號裏的字詞為粵語中的另一種講法。

③ 如何讓孩子玩互動主題遊戲

❶ 點選遊戲圖示，聆聽題目，例如：

❷ 按指示使用點讀筆點選正確的位置。

❸ 用點讀筆點出正確的答案。

🌟 每個主題均設有遊戲，題目隨機播放，答對便會繼續問下一題。答錯時，會重播問題。每題均設三次回答機會。當第三次仍答錯，便會問下一題。問題播放完畢，遊戲會自動結束。再次點選遊戲圖示，便可重新再玩。

🌟 透過點選語言圖示，便可選擇以哪種語言來玩遊戲。

🌟 在遊戲過程中，如想終止遊戲，可點選 。

4 如何讓孩子玩拼字遊戲

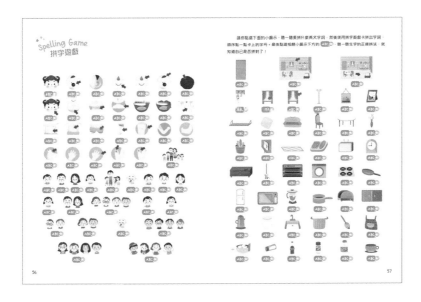

❶ 點選拼字遊戲 (P.56-67) 上任何一個小圖示，聽一聽要拼什麼英文字詞。例如：

❷ 順序排列拼字遊戲卡，拼出英文字詞。

❸ 用點讀筆順序點一點卡上的英文字母。

❹ 最後點選相關小圖示下方的 ABC，聽一聽生字的正確拼法，就知道自己是否拼對了！

❺ 重複步驟 ❶ - ❹，嘗試拼出其他英文字詞。

My Body
我的身體

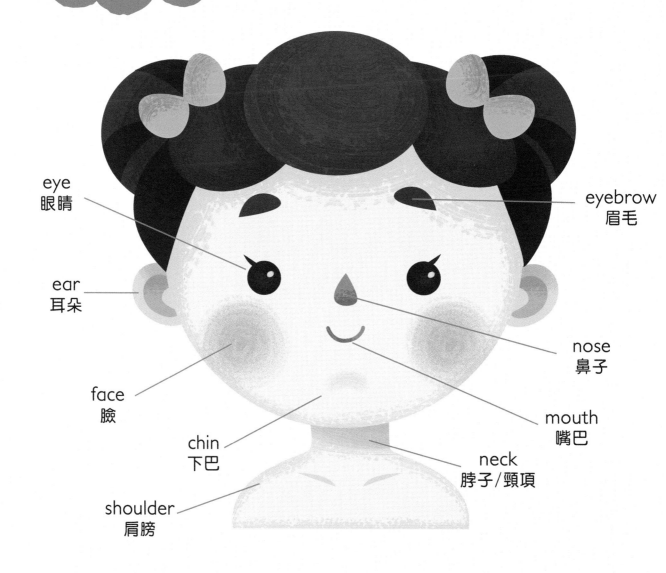

eye
眼睛

eyebrow
眉毛

ear
耳朵

nose
鼻子

face
臉

mouth
嘴巴

chin
下巴

neck
脖子/頸項

shoulder
肩膀

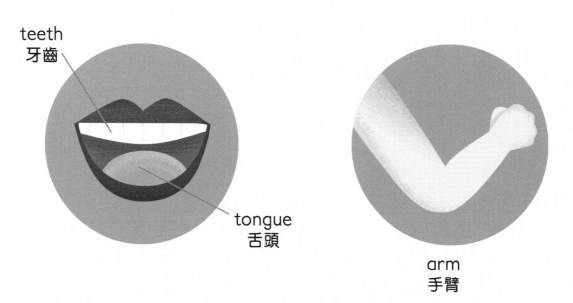

teeth
牙齒

tongue
舌頭

arm
手臂

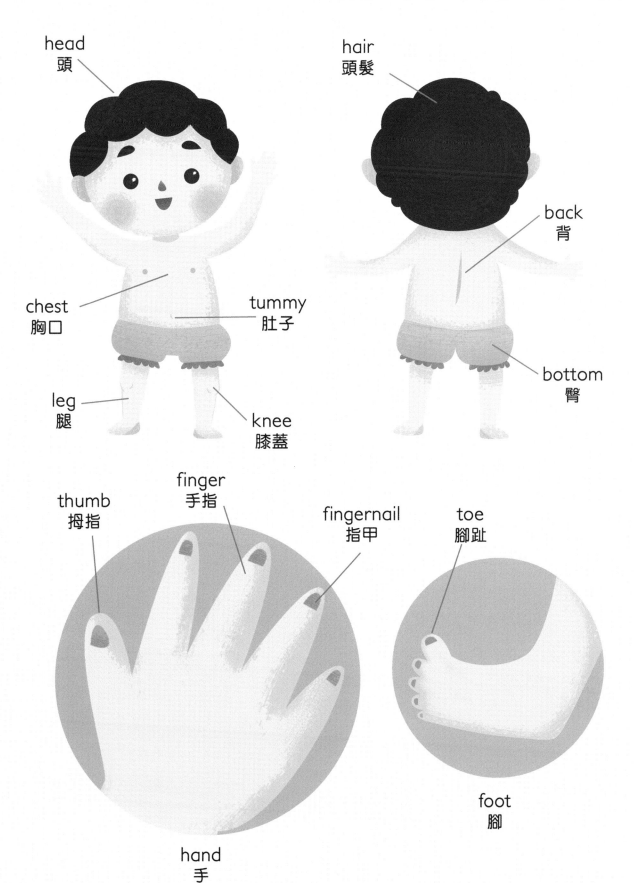

head
頭

hair
頭髮

back
背

chest
胸口

tummy
肚子

bottom
臀

leg
腿

knee
膝蓋

thumb
拇指

finger
手指

fingernail
指甲

toe
腳趾

hand
手

foot
腳

family 家庭

father 爸爸　　mother 媽媽

baby 嬰兒

daughter 女兒

son 兒子　　son 兒子　　daughter 女兒

grandparents 祖父母

grandmother 祖母（嫲嫲）

grandfather 祖父（爺爺）

parent 父母

 aunt 伯母/嬸嬸

 uncle 伯父/叔叔

 aunt 姑母（姑媽/姑姐）

 uncle 姑丈

 father 爸爸

 cousin 堂兄弟姐妹（堂兄弟姊妹）

me 我　　elder brother 哥哥

siblings
兄弟姐妹（兄弟姊妹）

elder brother
哥哥

elder sister
姐姐（姊姊）

grandmother
祖母（嫲嫲）

grandfather
祖父（爺爺）

younger brother
弟弟

younger sister
妹妹

grandson
孫兒

granddaughter
孫女

grandparents
外祖父母

grandmother
外祖母（婆婆）

grandfather
外祖父（公公）

mother
媽媽

aunt
姨母／姨（姨媽／阿姨）

uncle
姨丈

aunt
舅母

uncle
舅舅（舅父）

younger sister
妹妹

elder sister
姐姐（姊姊）

younger brother
弟弟

cousin
表兄弟姐妹（表兄弟姊妹）

9

In the Living Room
客廳裏

wall
牆壁

door
門

window
窗

clock
時鐘

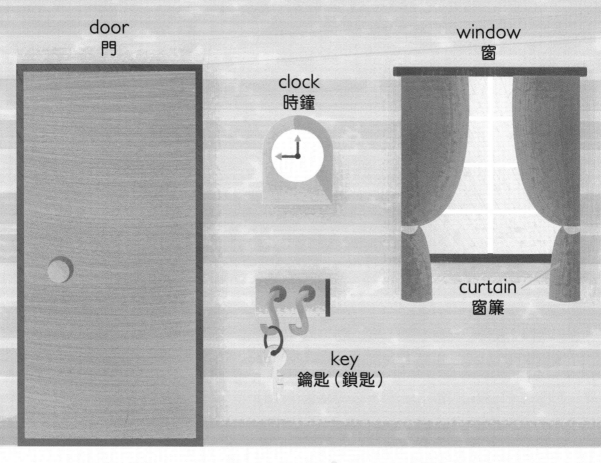

curtain
窗簾

key
鑰匙（鎖匙）

plant
盆栽

vase
花瓶

table
桌子

chair
椅子

floor
地面

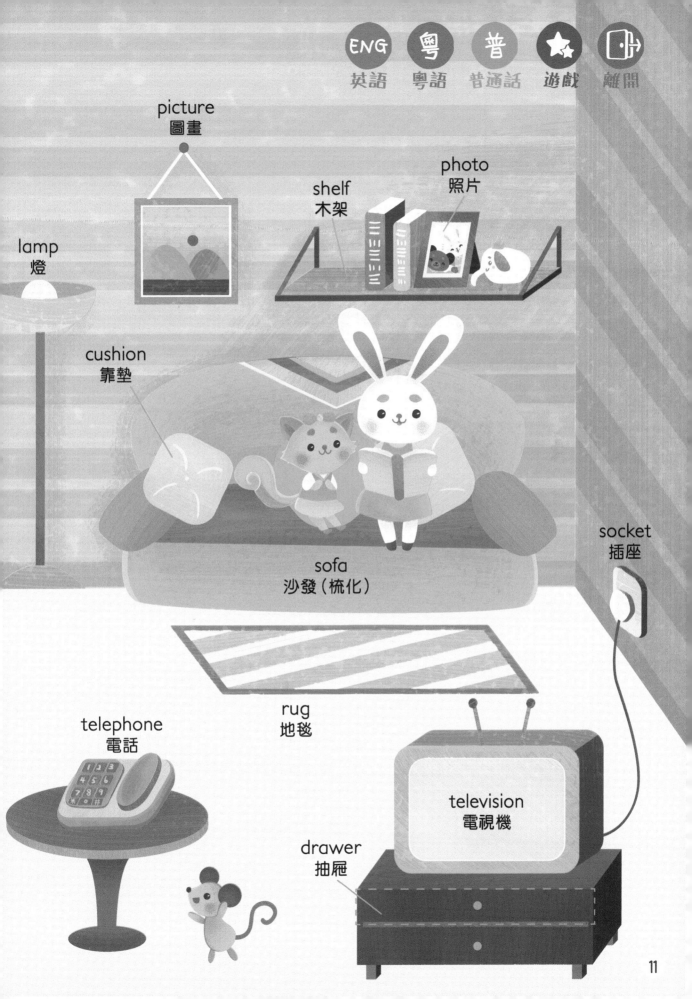

picture
圖畫

shelf
木架

photo
照片

lamp
燈

cushion
靠墊

socket
插座

sofa
沙發（梳化）

rug
地毯

telephone
電話

drawer
抽屜

television
電視機

11

In the Kitchen
廚房中

microwave oven
微波爐

toaster
烤麵包機（多士炉

fridge
冰箱（雪櫃）

kettle
燒水壺

mop
拖把

sink
洗碗槽

glass
玻璃杯

cloth
抹布

washing machine
洗衣機

whisk
手動打蛋器

mixer
攪拌機

bowl
碗

jar
廣口瓶

plate
碟子

fork
叉子

ENG 英語　粵 粵語　普 普通話　★ 遊戲　⬚ 離開

pepper
胡椒粉

soya sauce
醬油

oil
油

cupboard
櫥櫃

saucepan
深平底鍋

stove
爐灶

salt
鹽

frying pan
平底煎鍋

apron
圍裙

oven
烤箱（焗爐）

spoon
勺子（匙羹）

knife
刀

chopsticks
筷子

cup
杯子

sugar
糖

Bedroom and Bathroom
睡房和浴室

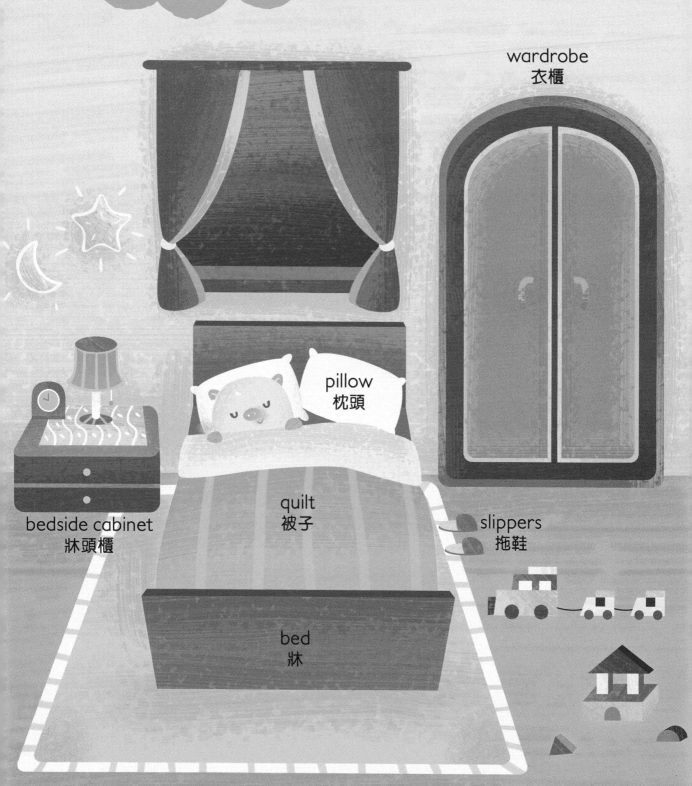

wardrobe
衣櫃

pillow
枕頭

quilt
被子

slippers
拖鞋

bedside cabinet
牀頭櫃

bed
牀

mirror
鏡子

hairbrush
梳子

toothbrush
牙刷

toothpaste
牙膏

towel
毛巾

tap
水龍頭

washbasin
洗臉盆

toilet
馬桶（廁所）

toilet paper
衞生紙（廁紙）

bath
浴缸

shampoo
洗髮露

soap
肥皂

My School
我的學校

poster
海報

classroom
課室

computer
電腦

mouse
滑鼠

school bag
書包

keyboard
鍵盤

classmate
同學

bookshelf
圖書架

storybook
故事書

uniform
校服

play mat
遊戲墊

whiteboard
白板

letter
字母

A B C

teacher
老師

student
學生

ruler
尺子

crayon
蠟筆

pencil
鉛筆

paint
顏料

eraser
橡皮（擦膠）

toy
玩具

picture card
圖卡

tray
托盤

17

see
看

hear
聽

touch
觸摸

laugh
笑

jump
跳

run
跑

sing
唱歌

clap
拍手

read
閱讀

smell
嗅

eat
吃

drink
喝

talk
談話

walk
步行

dance
跳舞

draw
畫圖畫

write
書寫

swim
游泳

My Toys
我的玩具

board game
桌上遊戲

toy train
玩具火車

building blocks
積木

yo-yo
溜溜球（搖搖）

doll
洋娃娃（公仔）

kite
風箏

marble
彈珠（波子）

toy rocket
玩具火箭

teddy bear
玩具熊

tea set
茶具玩具

toy box
玩具箱

ball
皮球

modelling clay
彩泥（泥膠）

toy car
玩具車

robot
機器人（機械人）

toy spaceship
玩具太空船

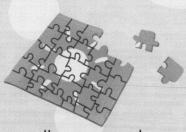

jigsaw puzzle
拼圖

rocking horse
木馬

Food and Drinks
食物和飲品

Food 食物

cereal
麥片

tomato
番茄

cheese
奶酪（芝士）

salad
沙拉（沙律）

egg
蛋

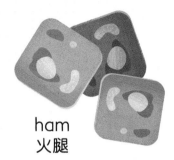
ham
火腿

bread
麵包

biscuit
餅乾

chocolate
巧克力（朱古力）

jelly
果凍（啫喱）

cake
蛋糕

ice cream
冰淇淋（雪糕）

noodles
麵條

rice
米飯

spaghetti
意大利麵

fish
魚

sandwich
三明治（三文治）

hamburger
漢堡包

soup
湯

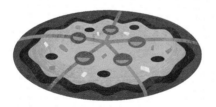

pizza
薄餅/比薩

chicken
雞

Drinks 飲品

water
開水

juice
果汁

coke
汽水

tea
茶

milk
牛奶

Vegetables and Fruit
蔬菜和水果

Vegetables 蔬菜

cabbage
捲心菜（椰菜）

pumpkin
南瓜

peas
豌豆

lettuce
生菜

broccoli
西蘭花/青花菜

onion
洋葱

mushroom
蘑菇

potato
馬鈴薯

carrot
胡蘿蔔（紅蘿蔔）

cucumber
黃瓜（青瓜）

corn
玉米（粟米）

Fruit 水果

orange
橙

pear
梨子

grape
葡萄（提子）

strawberry
草莓（士多啤梨）

mango
芒果

pineapple
菠蘿／鳳梨

lemon
檸檬

apple
蘋果

cherry
櫻桃（車厘子）

watermelon
西瓜

banana
香蕉

vest
內衣

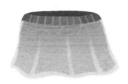

skirt
裙子

rain boots
雨鞋

raincoat
雨衣

underpants
內褲

swimsuit
游泳衣

dress
連衣裙

socks
襪子

belt
腰帶（皮帶）

sandals
涼鞋

tights
緊身褲

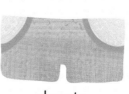

shorts
短褲

trousers
褲子

shoes
鞋子

mittens
連指手套

woolly hul
羊毛帽

scarf
圍巾（頸巾）

cap
鴨舌帽/便帽

swimming trunks
游泳褲

pyjamas
睡衣

jacket
短上衣

jeans
牛仔褲

boots
靴子

hat
帽子

coat
外套

T-shirt
短袖汗衫（T-恤）

27

Jobs
不同的職業

fireman
消防員

nurse
護士

doctor
醫生

policeman
警察

lawyer
律師

dentist
牙醫

teacher
老師

librarian
圖書館管理員

designer
設計師

chef
廚師

vet
獸醫

singer
歌星

hairdresser
理髮師

writer
作家

driver
司機

actor
演員

dancer
舞蹈演員（舞蹈員）

astronaut
宇航員（太空人）

scientist
科學家

cleaner
清潔員

artist
藝術家

waitress
侍應生

athlete
運動員

journalist
記者

Things That Go
交通工具

helicopter
直升機

car
汽車

wheel
輪子

bus
公共汽車（巴士）

motorcycle
摩托車（電單車）

taxi
計程車（的士）

police car
警車

van
客貨車

fire engine
消防車

ambulance
救護車

ship
輪船

boat
小船

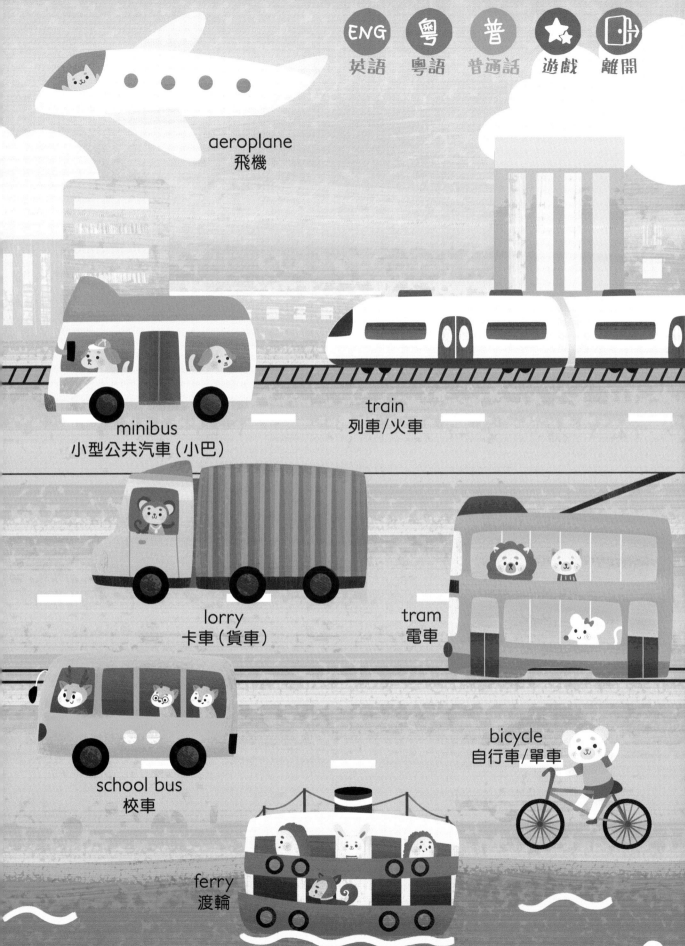

aeroplane
飛機

minibus
小型公共汽車（小巴）

train
列車/火車

lorry
卡車（貨車）

tram
電車

school bus
校車

bicycle
自行車/單車

ferry
渡輪

In the City
城市裏

park
公園

school
學校

bookshop
書店

café
咖啡室

bus stop
公車站（巴士站）

bakery
麵包店

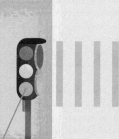

traffic light
交通信號燈/紅綠燈

police station
警察局

post office
郵政局

church
教堂

road
道路

bank
銀行

subway
行人隧道

supermarket
超級市場

mall
購物中心

cinema
電影院/戲院

Restaurant

restaurant
餐廳

museum
博物館

STATION

station
車站

zebra crossing
斑馬線

fire station
消防局

hospital
醫院

footbridge
行人天橋

HOTEL

hotel
酒店

At the Park
公園裏

sky
天空

sun
太陽

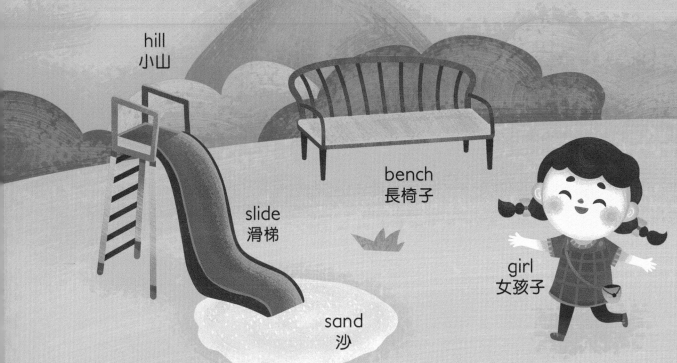

hill
小山

bench
長椅子

slide
滑梯

girl
女孩子

sand
沙

boy
男孩子

butterfly
蝴蝶

fountain
噴泉

bush
草叢

flower
花

bee
蜜蜂

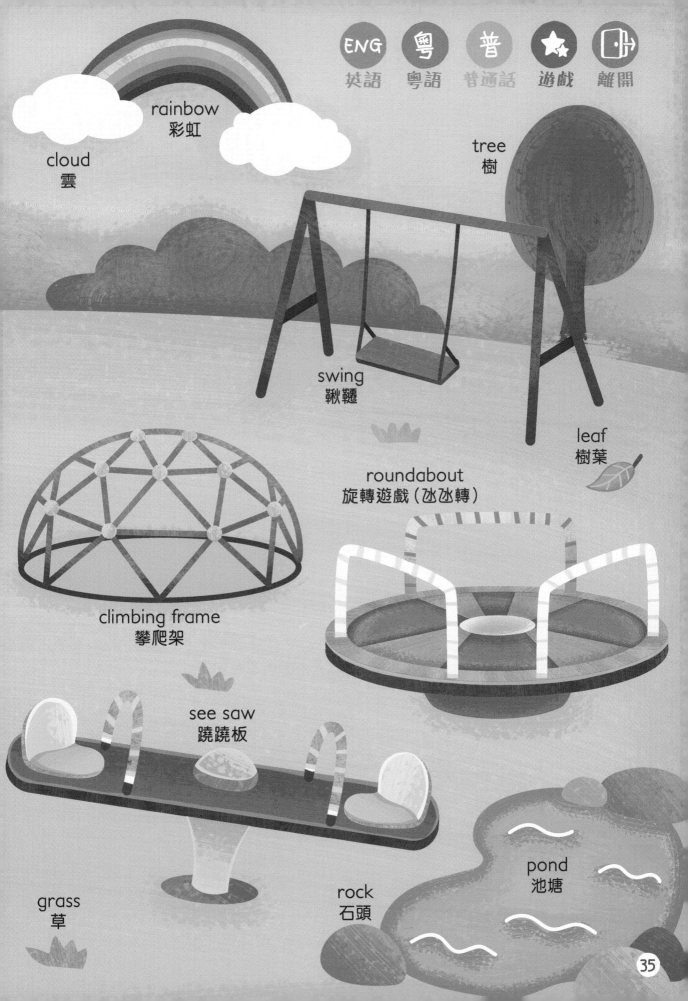

rainbow
彩虹

cloud
雲

tree
樹

swing
鞦韆

leaf
樹葉

roundabout
旋轉遊戲（氹氹轉）

climbing frame
攀爬架

see saw
蹺蹺板

grass
草

rock
石頭

pond
池塘

35

My Pets
我的寵物

tortoise
烏龜

cat
貓

kitten
小貓

puppy
小狗

dog
狗

goldfish
金魚

rabbit
兔

hamster
倉鼠

guinea pig
天竺鼠

bird
鳥

On the Farm
農場裏

hen
母雞

chick
小雞

duck
鴨

piglet
小豬

duckling
小鴨

foal
小馬

horse
馬

pig
豬

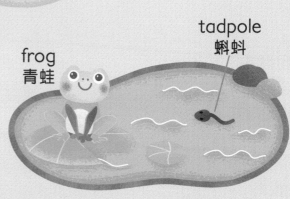

tadpole
蝌蚪

frog
青蛙

sheep
綿羊

lamb
小綿羊

cow
乳牛/奶牛

goat
山羊

calf
小牛

kid
小山羊

At the Zoo
動物園裏

polar bear
北極熊

lion
獅子

tiger
老虎

flamingo
紅鶴

owl
貓頭鷹

gorilla
大猩猩

snake
蛇

giraffe
長頸鹿

zebra
斑馬

crocodile
鱷魚

whale
鯨魚

shark
鯊魚

panda
熊貓

bear
熊

parrot
鸚鵡

koala
樹熊／無尾熊

kangaroo
袋鼠

elephant
大象

monkey
猴子

hippo
河馬

ostrich
鴕鳥

sea turtle
海龜

octopus
章魚（八爪魚）

dolphin
海豚

seal
海豹

penguin
企鵝

39

Months, Days and Time
月份、日和時間

calendar 月曆

Days
日

Monday 星期一	Tuesday 星期二	Wednesday 星期三	Thursday 星期四	Friday 星期五	Saturday 星期六	Sunday 星期日
	1	2	3	4	5	6
7	8	9	10	11	12	13
14	15	16	17	18	19	20
21	22	23	24	25	26	27
28	29	30	31			

week
一星期/一周

Months 月份

January 一月	February 二月	March 三月	April 四月	May 五月	June 六月
1	2	3	4	5	6
7	8	9	10	11	12
July 七月	August 八月	September 九月	October 十月	November 十一月	December 十二月

Time 時間

o'clock
……時正

half past
……時半/時三十分

a quarter past
……時十五分

a quarter to
……時四十五分

hour 時 —  — second 秒

minute 分

morning
早上

afternoon
下午

evening
傍晚

night
晚上

Seasons and Weather
季節和天氣

Seasons
季節

spring
春天

umbrella
雨傘

summer
夏天

autumn
秋天

winter
冬天

Weather
天氣

sunny
天晴的

rainy
下雨的

snowy
下雪的

typhoon
颱風

lightning
閃電

thunderstorm
雷暴

foggy
有霧的

cloudy
多雲的

windy
大風的

warm
和暖的

hot
炎熱的

cool
清涼的

cold
寒冷的

Numbers
數字

one
一

two
二

three
三

four
四

five
五

six
六

seven
七

eight
八

nine
九

ten
十

eleven
十一
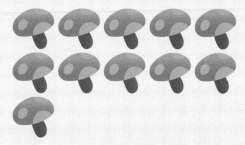

twelve
十二

thirteen
十三

fourteen
十四

fifteen
十五

44

Colours and Shapes
顏色和形狀

ENG 英語　粵 粵語　普 普通話　遊戲　離開

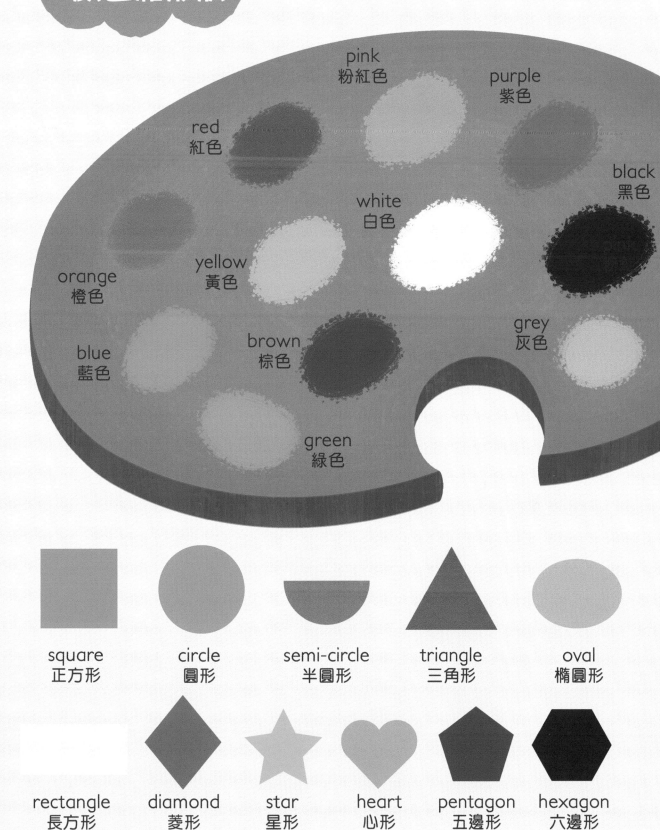

pink 粉紅色

purple 紫色

red 紅色

black 黑色

white 白色

yellow 黃色

orange 橙色

grey 灰色

blue 藍色

brown 棕色

green 綠色

square 正方形

circle 圓形

semi-circle 半圓形

triangle 三角形

oval 橢圓形

rectangle 長方形

diamond 菱形

star 星形

heart 心形

pentagon 五邊形

hexagon 六邊形

45

Adjective
形容詞

noisy
吵鬧的

quiet
文靜的

strong
強壯的

curly
捲曲的

straight
直的

naughty
淘氣的

angry
生氣的

tired
疲倦的

clean
乾淨的

dirty
骯髒的

happy
快樂的

sad
傷心的

old
年老的

young
年青的

thirsty
口渴的

lazy
懶惰的

busy
忙碌的

hungry
飢餓的

Opposite
相反詞

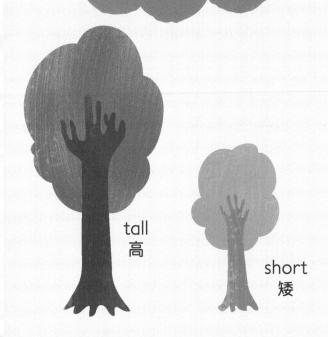

tall
高

short
矮

inside
在……裏面

outside
在……外面

front
正面

back
背面

fat
胖

thin
瘦

big
大

small
小

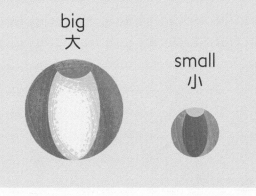

up
向上

down
向下

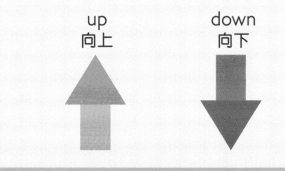

long
長

short
短

on
在……上面

under
在……下面

fast
快

slow
慢

dry
乾

wet
濕

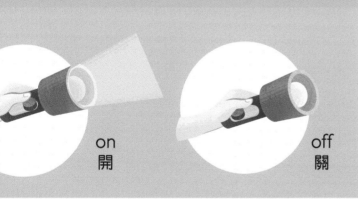

on
開

off
關

heavy
重

light
輕

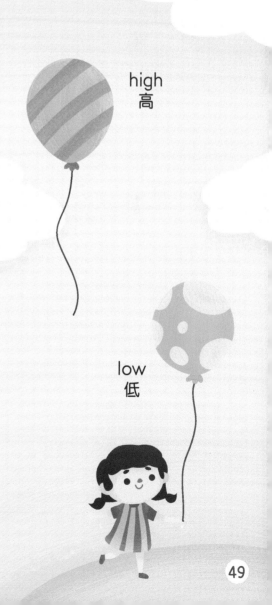

high
高

low
低

49

Festivals
節日

red banner
春聯（揮春）

red packet
紅包（利是）

rice cake
年糕

fireworks
煙花

candy box
攢盒

lion dance
舞獅

peach blossom
桃花

Mid-Autumn Festival 中秋節

moon
月亮

moon cake
月餅

star fruit
楊桃

lantern
燈籠

Christmas 聖誕節

Merry Christmas!
聖誕快樂！

Ho Ho

Santa Claus
聖誕老人

Christmas tree
聖誕樹

turkey
火雞

Christmas stocking
聖誕襪

gingerbread
薑餅

present
禮物

Easter 復活節

Easter egg
復活蛋

Easter bunny
復活兔

basket
籃子

Dragon Boat Festival 端午節

dragon boat
龍舟

rice dumpling
粽子

Musical Instruments
樂器

ENG 英語　粵 粵語　普 普通話　★ 遊戲　離開

drum
鼓

xylophone
木琴

guitar
吉他（結他）

flute
長笛

harp
豎琴

violin
小提琴

trumpet
小號

piano
鋼琴

triangle
三角鐵

recorder
豎笛（牧童笛）

recycling bin
回收箱

| paper 紙張 | plastic 塑料 | metal 金屬 |

plastic bottle
塑料瓶

magazine
雜誌

can
罐子

newspaper
報紙

clothes
衣物

battery
電池

Space and Landform
太空和地貌

space　太空

Mercury
水星

Venus
金星

Earth
地球

Mars
火星

volcano
火山

valley
山谷

plain
平原

island
島嶼

lake
湖泊

river
河流

Jupiter
木星

Saturn
土星

Uranus
天王星

Neptune
海王星

mountain
山

waterfall
瀑布

desert
沙漠

ocean
海洋

Spelling Game
拼字遊戲

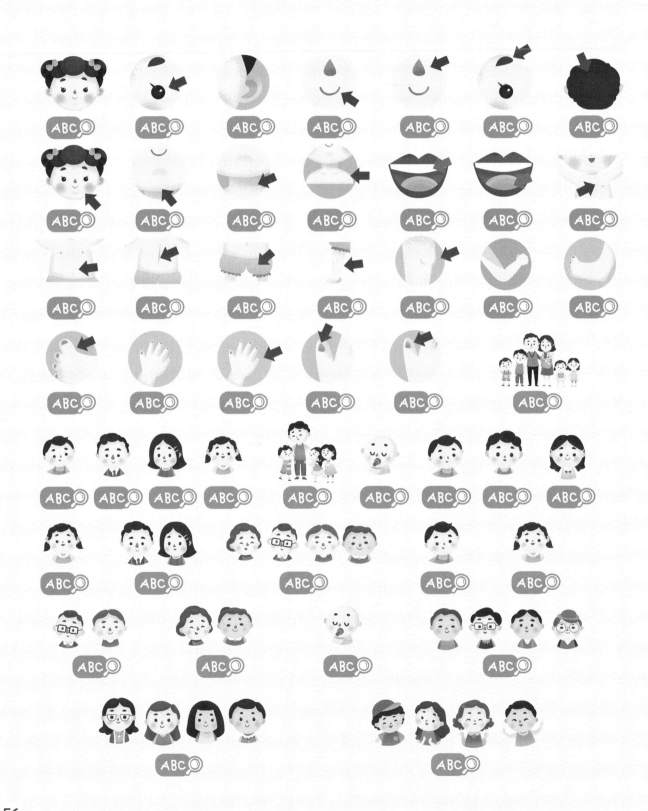

請你點選下面的小圖示，聽一聽要拼什麼英文字詞，然後使用拼字遊戲卡拼出字詞，順序點一點卡上的字母。最後點選相關小圖示下方的 ABC，聽一聽生字的正確拼法，就知道自己是否拼對了！

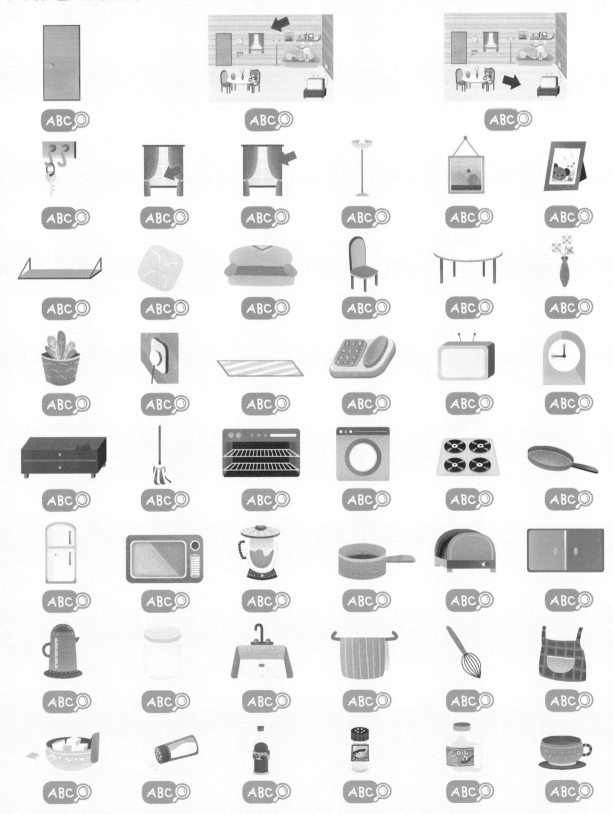

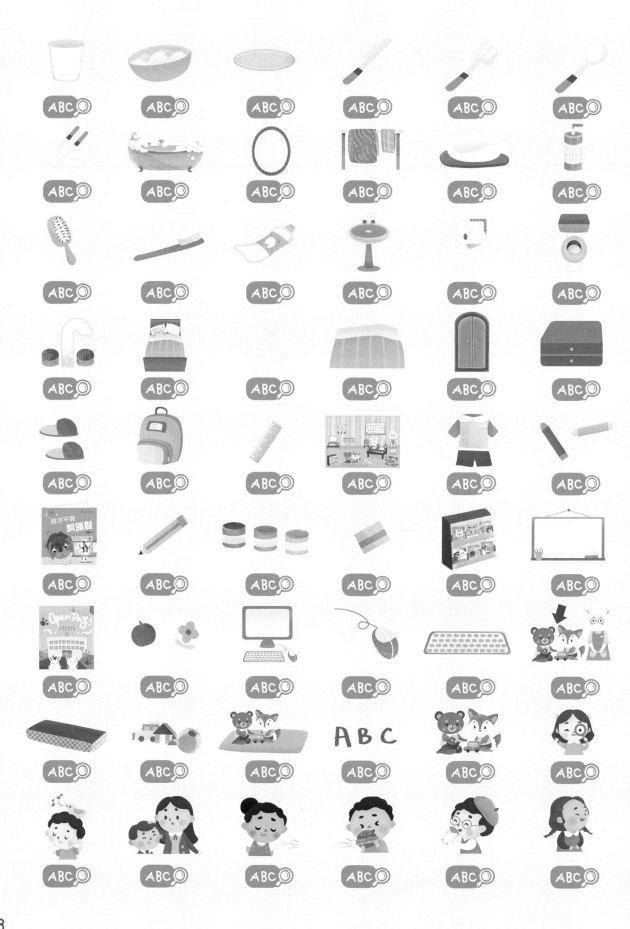

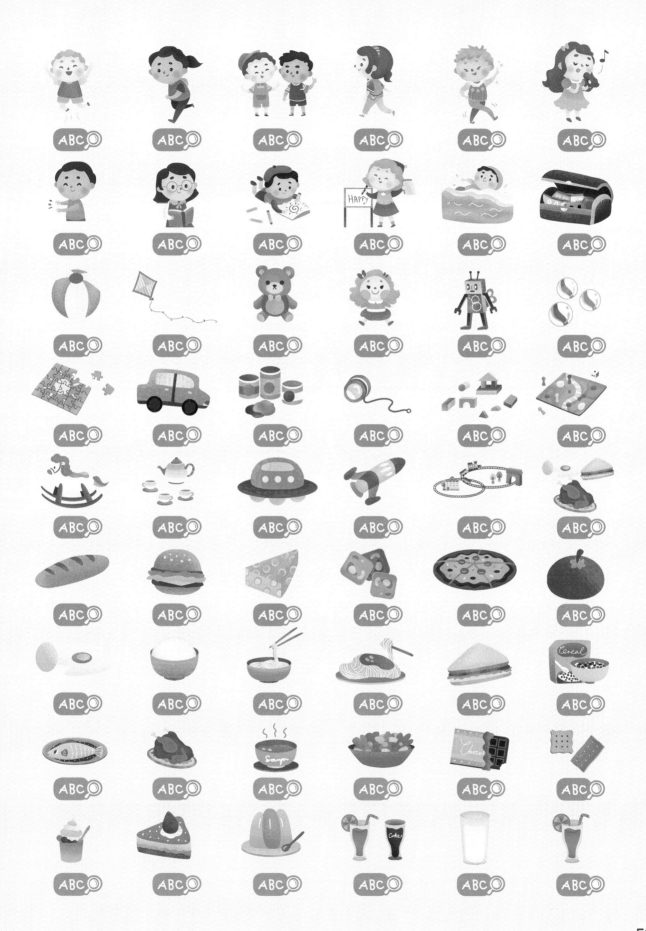

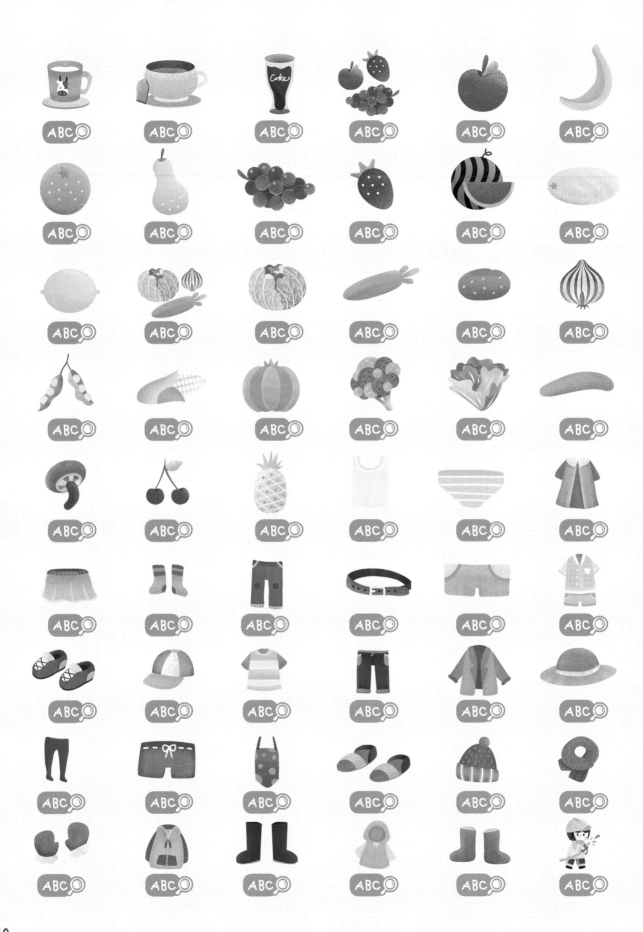

ABC
ABC
ABC
ABC
ABC
ABC

ABC
ABC
ABC
ABC
ABC
ABC

ABC
ABC
ABC
ABC
ABC
ABC

ABC
ABC
ABC
ABC
ABC
ABC

ABC
ABC
ABC
ABC
ABC
ABC

ABC
ABC
ABC
ABC
ABC
ABC

ABC
ABC
ABC
ABC
ABC
ABC

ABC
ABC
ABC
ABC
ABC
ABC

ABC
ABC
ABC
ABC
ABC
ABC

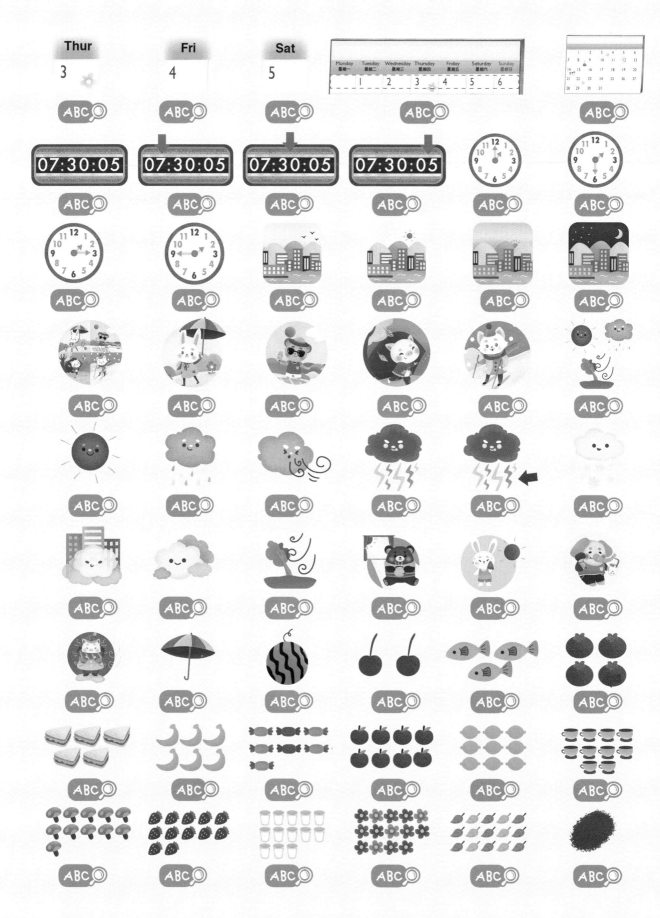

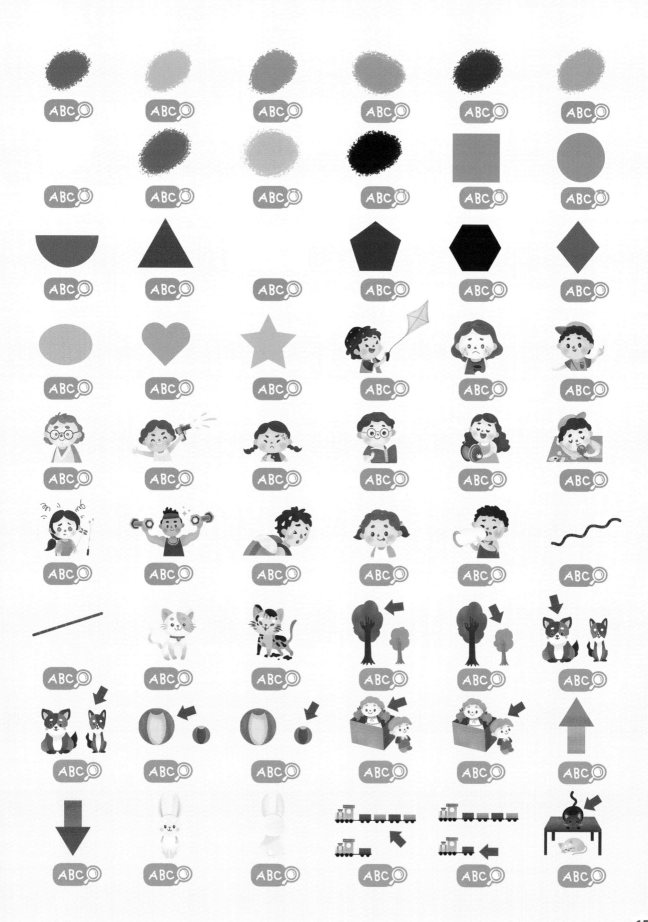

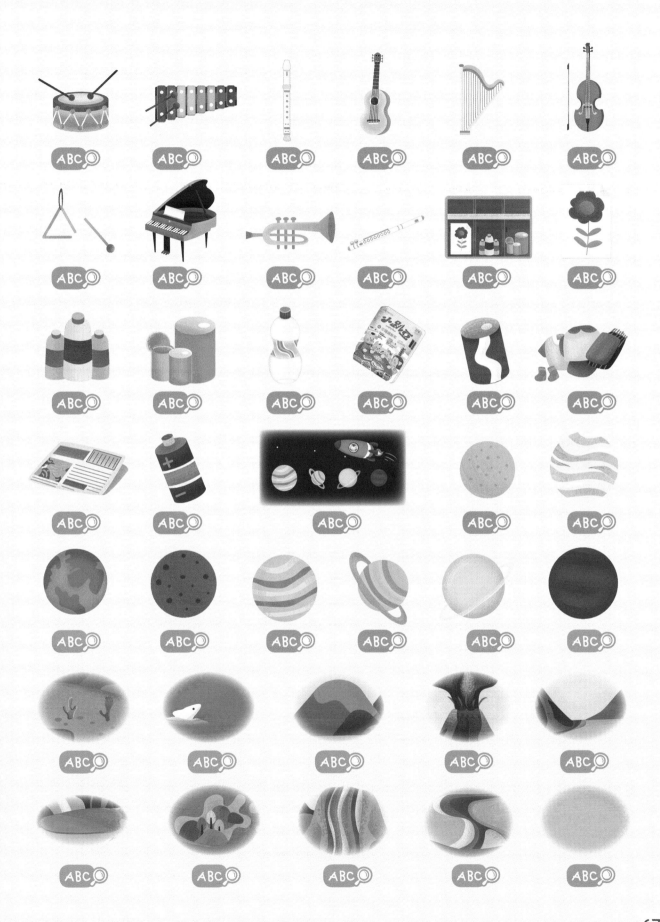

粵語

普通話

Dictionary
小字典

*小字典收錄了本圖典內的字詞。

Aa	Bb	Cc	Dd	Ee	Ff
Gg	Hh	Ii	Jj	Kk	Ll
Mm	Nn	Oo	Pp	Qq	Rr
Ss	Tt	Uu	Vv	Ww	Xx
Yy	Zz				

輸入 Enter	退格 Backspace	刪除 Delete	離開 Exit

如何使用小字典

示範短片

❶ 點選 Dictionary 小字典，啟動小字典功能，然後點選粵語或普通話語言圖示。
（每次切換語言，請再點選 Dictionary 小字典）

❷ 順序點選英文字母，拼出你想查詢的字詞。

❸ 如按錯字母，可點選 退格 Backspace，刪除上一個輸入的字母，然後重新點選。

❹ 如要取消輸入的整個字詞，可點選 刪除 Delete。

❺ 輸入字詞後，點選 輸入 Enter，即可聽到字詞的英語和粵語或英語和普通話的發音。

❻ 如要退出小字典，請點選 離開 Exit。